OUT OF THIS WORLD

Meet NASA Inventor Saptarshi Bandyopadhyay and His Team's

Giant Lunar Telescope

WORLD BOOK

www.worldbook.com

World Book, Inc.
180 North LaSalle Street
Suite 900
Chicago, Illinois 60601
USA

For information about other World Book publications, visit our website at www.worldbook.com or call 1-800-WORLDBK (967-5325).

For information about sales to schools and libraries, call 1-800-975-3250 (United States), or 1-800-837-5365 (Canada).

© 2024 (print and e-book) by World Book, Inc. All rights reserved. No part of this publication may be reproduced, stored in a retrieval system, or transmitted in any form or by any means (electronic, mechanical, photocopying, recording, or otherwise) without written permission from World Book, Inc.

WORLD BOOK and the GLOBE DEVICE are registered trademarks or trademarks of World Book, Inc.

Produced in collaboration with the National Aeronautics and Space Administration (NASA).

Library of Congress Cataloging-in-Publication Data for this volume has been applied for.

Out of This World
ISBN: 978-0-7166-6564-9 (set, hc.)

Giant Lunar Telescope
ISBN: 978-0-7166-6567-0 (hc.)

Also available as:
ISBN: 978-0-7166-6575-5 (e-book)
ISBN: 978-0-7166-6583-0 (soft cover)

Staff

Editorial

Vice President
Tom Evans

Senior Manager, New Content
Jeff De La Rosa

Writer
William D. Adams

Editor
Emma Flickinger

Curriculum Designer
Caroline Davidson

Proofreader
Nathalie Strassheim

Indexer
Nathaniel Lindstrom

Graphics and Design

Senior Visual Communications Designer
Melanie Bender

Digital Asset Specialist
Rosalia Bledsoe

Acknowledgments

Cover	© Vladimir Vustyansky	20-21	© Artsiom P/Shutterstock
		22-25	NASA
3	© Artsiom P/Shutterstock; © Vladimir Vustyansky	26-27	© Vladimir Vustyansky
4-5	© J_Chaikom, Shutterstock	29	Saptarshi Bandyopadhyay
6-7	NASA	30-31	© Mark Williamson, Science Photo Library
8-9	NASA/STScI	32-33	Matthew Luem/ NASA
10-11	© Triff/Shutterstock	34-35	© Vladimir Vustyansky; WORLD BOOK
12-13	© Sarawut Sriphakdee, Shutterstock	36-37	NASA/JPL-Caltech/J.D. Gammell
14-15	Saptarshi Bandyopadhyay	38-39	Saptarshi Bandyopadhyay
16-17	© Marcel Clemens, Shutterstock	40-43	© Vladimir Vustyansky
18-19	© Xinhua/Alamy Images	44	Saptarshi Bandyopadhyay

Contents

- 4 Introduction
- 8 Looking back in time
- 10 The Dark Ages
- 12 Dark matter and dark energy
- 14 INVENTOR FEATURE: Seems like destiny
- 16 Radio astronomy
- 18 Radio telescopes
- 20 Light from the Dark Ages
- 22 In the shadow of the ionosphere
- 24 BIG IDEA: A moon-sized shield
- 28 INVENTOR FEATURE: Student satellite project
- 30 BIG IDEA: Reflector mesh
- 32 Deployable structures
- 34 BIG IDEA: Harpoons and weights
- 36 INVENTOR FEATURE: Keep innovating
- 38 INVENTOR FEATURE: Photography and national parks
- 40 Lunar challenges
- 42 A lunar renaissance
- 44 Saptarshi Bandyopadhyay and Haripriya Vaidehi Narayanan
- 45 Glossary
- 46 Review and reflect
- 48 Index

Glossary There is a glossary of terms on page 45. Terms defined in the glossary are in boldface type that **looks like this** on their first appearance on any spread (two facing pages).

Pronunciations (how to say words) are given in parentheses the first time some difficult words appear in the book. They look like this: pronunciation (pruh NUHN see AY shuhn).

Introduction

Have you ever looked up at the vast night sky? Every bit of space—even the seemingly dark reaches between stars—harbors hidden clues to the universe's history. One of the most important mysteries these clues could unravel is why matter clumped together to form gas clouds, stars, and galaxies, instead of spreading out into a featureless sea of evenly spaced **atoms.**

Scientists called **astrophysicists** need colossal telescopes to gather these clues. But, the signals they seek are extremely faint and easily blocked by Earth's **atmosphere.**

Craters on the moon

Engineer Saptarshi Bandyopadhyay (BAHN dyoh PAHD yeh) wants to build a telescope to gather these signals far from the reach of Earth's **atmosphere**—the Lunar Crater Radio Telescope (LCRT). But how do you build a giant telescope on the moon?

The NASA Innovative Advanced Concepts program. The titles in the *Out of This World* series feature projects that have won grant money from a group formed by the United States National Aeronautics and Space Administration, or NASA. The NASA Innovative Advanced Concepts program (NIAC) provides funding to teams working to develop bold new advances in space technology. You can visit NIAC's website at www.nasa.gov/niac.

Meet Saptarshi Bandyopadhyay.

" I'm an engineer at the Jet **Propulsion** Laboratory (JPL) in California. I'm developing a mission design that will vastly improve our understanding of the universe's history, giving support to—or perhaps casting doubt upon—some of the grandest theories in astrophysics. "

Looking back in time

Light travels fast—some 186,282 miles (299,792 kilometers) per second! But, the universe is mind-bogglingly large. The time it takes for light to reach Earth from far-flung places adds up. For example, light from the sun takes eight minutes to reach Earth.

The next-closest stars are so far away that the distance is measured in *light-years*. A light-year is the distance light travels in one Earth year, about 5.88 trillion miles (9.46 trillion kilometers). The nearest star, Proxima Centauri, is a little over four light-years from Earth. When we look at this star, we are really seeing what it looked like four years ago. Sirius, the brightest star in the night sky, is 8.6 light-years away. Other stars are dozens, hundreds, or even thousands of light-years away. And, these are only stars within our own galaxy, the Milky Way. The Andromeda Galaxy, our nearest major galactic neighbor, is about 2.5 million light-years away. Further galaxies measure many millions or even billions of light-years away.

The constant, *finite* (limited) speed of light means that observations of far-away objects actually provide snapshots of the universe's history. **Astrophysicists** can use them to look back in time and learn how the universe formed.

The Dark Ages

❚❚ Understanding the evolution of the universe is one of the most fundamental questions in science. ❚❚ —Saptarshi

From studying these "snapshots" of the early universe and other evidence, astrophysicists think that the universe as we know it expanded rapidly about 13.8 billion years ago. This event is called the **big bang.** At the time, the universe was hot and dense beyond measure. As the universe spread out and cooled, matter emerged in the form of hydrogen atoms. For many tens of thousands of years, the universe was a featureless ocean of hydrogen atoms. Then, the hydrogen began to clump together and ignite, forming the first stars. The first wave of star formation happened some 13.5 billion years ago.

Astrophysicists call the period before the stars formed—and after the big bang—the Dark Ages. What happened during the Dark Ages? Why did hydrogen clump together at all? And, why did it clump up when it did? Scientists suspect that the answers have to do with two major ingredients in the universe that we know very little about. These mysterious ingredients are called **dark matter** and **dark energy.**

Dark matter and dark energy

Dark matter is the invisible substance or substances thought to make up the majority of the matter in the universe. Unlike ordinary matter, dark matter does not give off, reflect, or absorb light. Scientists have detected dark matter through the effects of its **gravitational pull** on objects that we can see. Their observations suggest that about 85 percent of the matter in the universe is dark matter.

The universe has continued to expand after the big bang. Judging by the amount of matter, astrophysicists would expect the universe to stop expanding and eventually even start shrinking under its gravitational pull. Instead, its expansion is *accelerating* (speeding up). Astrophysicists have named the force responsible for this acceleration dark energy.

Scientists think dark matter and dark energy play a major role in the evolution of the universe. But they know very little about either. Saptarshi's LCRT could help to untangle these mysteries.

Inventor feature:
Seems like destiny

Sometimes, a person's first or last name suits their profession. People might joke that the name sealed the person's destiny—guiding the person to that profession. In Saptarshi's case, there is some truth to the idea!

> *Saptarshi* is the name for the Big Dipper constellation in Sanskrit, the ancient language of India. I don't know how much my parents knew about the name's astronomical importance, but I was very excited about the name and its meaning and its astronomical significance from a very young age. —Saptarshi

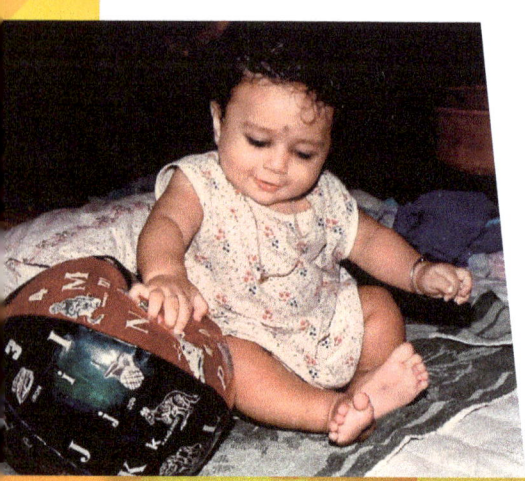

Saptarshi was looking at constellations by the time he was 3. His parents used tennis balls and candlelight to teach him about the **solar system,** day-and-night cycles, seasons, and eclipses. They also encouraged his aptitude for math, which is the foundation of astronomy and engineering.

Saptarshi maintained an interest in science and technology through school. He was not just interested in learning what scientists had already discovered. He wanted to understand what wasn't yet known and work on figuring it out.

> I have always been very fascinated with understanding what humans know. —Saptarshi

Saptarshi won a gold medal in the 2004 International Astronomy Olympiad (IAO). The IAO is a competition in which teams of students from different countries test their astronomical knowledge. Saptarshi's participation granted him access to research observatories in India. One of the facilities was the Giant Metrewave Radio Telescope (GMRT), outside Pune. At GMRT, Saptarshi began to understand that more powerful telescopes were needed.

> That was when I really understood this concept that there are these **wavelengths** of light that you can't see—of 10 meters (30 feet) or longer—that no one knows anything about. —Saptarshi

Radio astronomy

The GMRT that Saptarshi got to visit is an important observatory in **radio astronomy.** Astronomical bodies of all kinds give off radio waves. Radio waves have wavelengths at least 1 millimeter (0.04 inches) in length. But, they can be far longer—many tens of yards or meters long. (Wavelength is the distance between two crests or two troughs of a wave, such as a light wave.)

A radio telescope works much the same way as other telescopes. Most radio telescopes collect radio waves with a large *primary* instrument, often called a dish antenna or simply a dish. The dish has the same shallow bowl shape as the mirror of a reflecting telescope. Radio waves, however, are much longer than light waves. So, the dish must be much larger in diameter to collect them.

The dish focuses the waves onto an antenna, which translates them into electric signals. The signals contain information on the intensity of waves at different wavelengths. A radio receiver amplifies the signals and sends them to a computer.

The computer analyzes the *spectrum* (range of wavelengths) of the radio waves or produces an image of their source.

One of the 30 dishes that comprise the Giant Metrewave Radio Telescope (GMRT) in India

❝You usually want dish antenna to be ten times the size of wavelength that you're interested in.❞ —Saptarshi

Radio telescopes

In addition to the GMRT, there are other radio telescopes that are colossal in size and have made big discoveries.

The radio telescope at the Arecibo Observatory in Puerto Rico was the world's most powerful when it opened in 1969. Its dish had a diameter of 305 meters (1,000 feet). Unfortunately, delayed maintenance and storm damage caused the telescope to collapse beyond repair in 2020.

The Green Bank Telescope in West Virginia is one of the world's largest movable structures. Completed in 2001, its dish antenna is more than 100 meters (330 feet) across.

The Five-hundred-meter Aperture Spherical radio Telescope (FAST) is a giant radio telescope in Guizhou Province, China. Completed in 2016, the telescope measures about 500 meters (1,600 feet) in diameter.

The FAST nestles within valley walls in Guizhou Province, China.

Light from the Dark Ages

Saptarshi's LCRT should enable us to peer into the dark ages of the universe—because they were not entirely dark. Once every half-a-billion years or so, a hydrogen **atom** *spontaneously* (naturally) emits a radio wave with an 21-centimeter (8-inch) wavelength. Such an emission from any one hydrogen atom is rare. But, the sheer number of free hydrogen atoms in the early universe would have resulted in a strong signal at this wavelength.

If the wavelength of the hydrogen emission remained at 21 centimeters, then a dish a little over 2 meters (6 ½ feet) in diameter should be able to pick it up. But, the universe has continued to expand after the big bang. As space expands, it stretches out the wavelengths of light— including radio waves— traveling through it. This effect is called cosmological *redshift*.

> ❞ Things far away in the early universe are also moving away from us at a very high velocity. Because of that, this 21-centimeter light gets stretched to 10 meters (30 feet) or even longer wavelengths. ❞ —Saptarshi

In the shadow of the ionosphere

The Arecibo and FAST radio telescopes have made great discoveries from right here on Earth. And, FAST is large enough to detect hydrogen emissions from the Dark Ages. So, why do we need a radio telescope on the moon?

> ❚❚ The super-ultra-long wavelengths—wavelengths 10 meters (30 feet) or longer—are not visible from Earth. Why? Because waves of these lengths interact with the Earth's **ionosphere** and get absorbed. ❚❚ —Saptarshi

The ionosphere is a part of Earth's **atmosphere** that has many electrically charged particles. All those charged particles block and scramble ultra-long wavelength radio waves before they reach the surface. So, Earth-bound observatories can only study shorter wavelengths.

In **orbit** above the ionosphere, the situation does not get much better. The sun also produces radio waves, so any orbiting observatory would have to be shielded from this interference. Even then, the ionosphere would still hinder observations.

❚❚ Outside Earth, like say on a satellite, the ionosphere is such a strong noise source that it will pollute that signal. ❚❚ —Saptarshi

The ionosphere also interacts with the solar wind to produce auroras, as seen from this photograph taken from the International Space Station.

Big idea:
A moon-sized shield

To look deeper into the early history of the universe, **radio astronomy** must get out from under the shadow of the **ionosphere**. Fortunately, there is a great location right next door to Earth.

The moon has several benefits as a site for a radio telescope. The most important advantage is that the moon lacks an ionosphere. There is no **atmosphere** to become *ionized* (electrically charged).

❚❚ We need to go to a place that is hidden from Earth's noises. And the best place that is close to us is the far side of the moon. The far side of the moon never sees Earth because the same side of the moon always faces our planet. You can think of the moon as a shield that protects the Lunar Crater Radio Telescope (LCRT) from all the noise coming from Earth. ❚❚ —Saptarshi

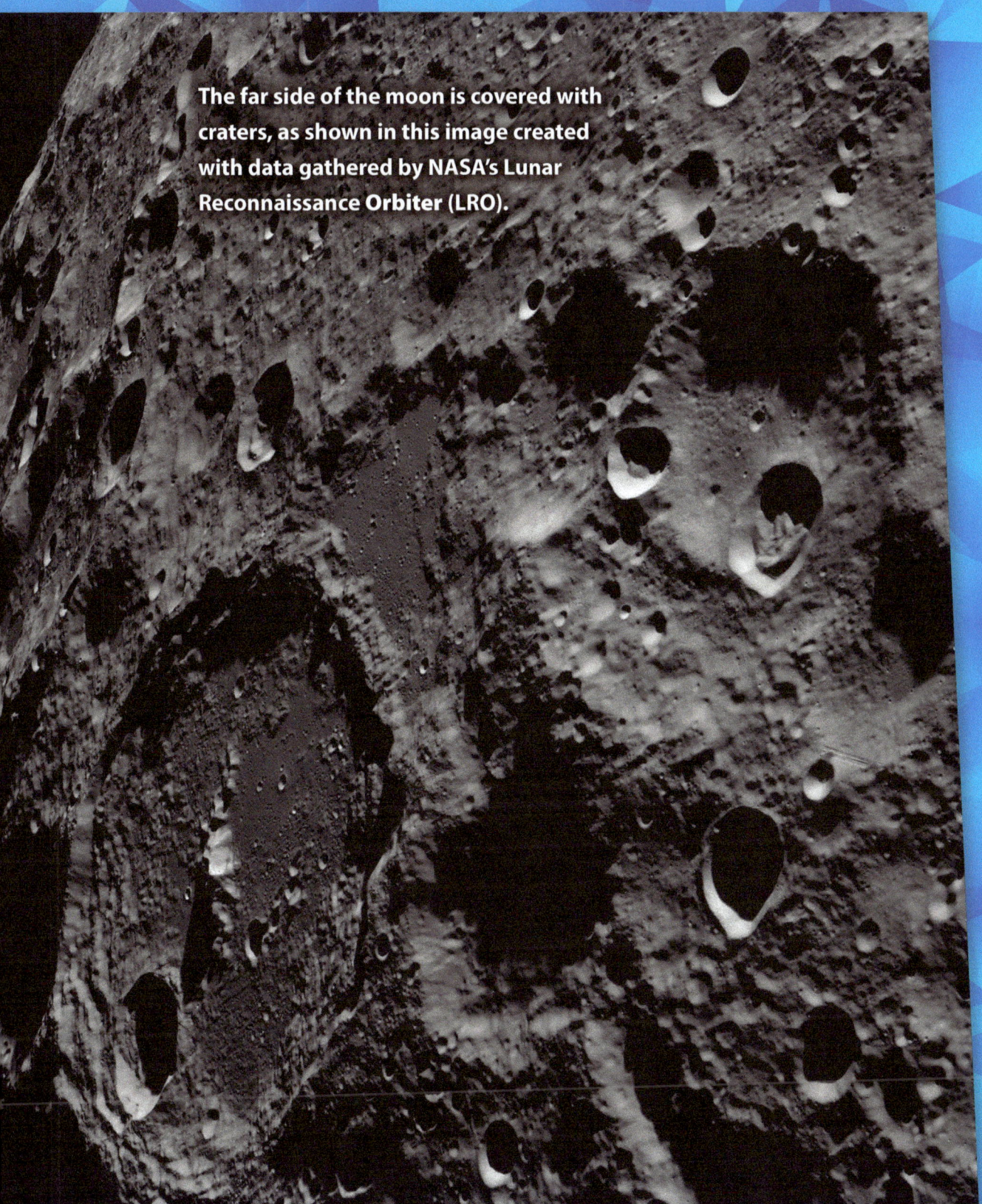
The far side of the moon is covered with craters, as shown in this image created with data gathered by NASA's Lunar Reconnaissance **Orbiter** (LRO).

Big idea:
A moon-sized shield cont.

A radio telescope also needs a colossal reflector dish to focus long wavelengths. **Engineers** designed Arecibo and FAST inside natural bowl-shaped depressions. These sites took advantage of the shape of the land to minimize the amount of bracing needed. The moon is littered with such depressions—craters!

Artist's impression of LCRT deployed on the far side of the moon

On Earth, radio telescopes require massive concrete pylons to be sunk into the ground to support their weighty reflectors. On the moon's surface, the force due to **gravity** is only about one-sixth that on Earth. Modest anchors would be able to support the entire structure.

Because of the moon's **orbit** and rotation, most of its surface spends just under 15 Earth days in lunar night, shaded from sunlight. LCRT would operate during this 15-day period, shielded from the sun's radio interference. The telescope would lie dormant during the 15-day lunar daytime.

Look on the bright side!

Sometimes people call the part of the moon facing away from us the "dark side" of the moon. But it is not any darker than the side we can see! As the moon orbits Earth, half of it is in sunlight, just like Earth. When there is a full moon, the far side is totally dark, but when there is a new moon, the far side is totally light.

Inventor feature:
Student satellite project

When Saptarshi was an undergraduate student at the Indian Institute of Technology (IIT) Bombay, he founded and led a student effort to build and launch a satellite!

The first student satellite program was created at California Polytechnical University, San Luis Obispo, in the early 2000's.

❝ This idea that university students could build a satellite and launch it was just making the rounds all over the academic world. ❞ —Saptarshi

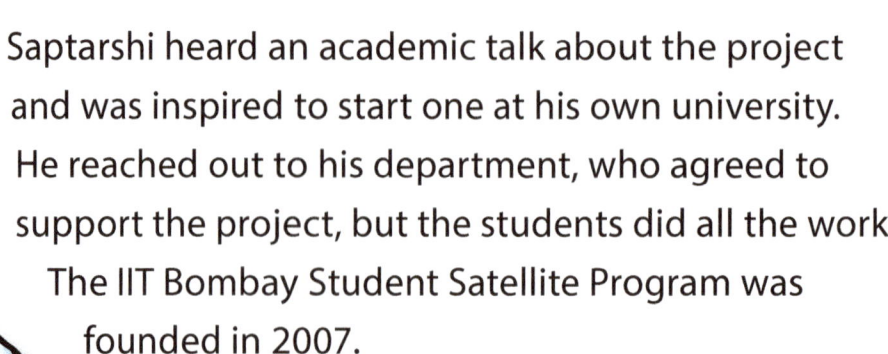

Saptarshi heard an academic talk about the project and was inspired to start one at his own university. He reached out to his department, who agreed to support the project, but the students did all the work. The IIT Bombay Student Satellite Program was founded in 2007.

The satellite Pratham, designed and built by the IIT Bombay Student Satellite Program, sits on display before it was launched in 2016.

Saptarshi completed his undergraduate work in 2010. But he stayed on for another year to ensure the project continued.

❝ One of the things we learned is the importance of continuity. People change, but the project has to keep moving forward. At every stage, people have to invest enough time so that when the next person comes to that position, they know why decisions were made five years ago at that level. ❞ —Saptarshi

In 2016, the satellite, named Pratham, was launched. Since then, the organization has developed improved components for small satellites, including an altitude **sensor** and antenna deployment system.

Big idea:
Reflector mesh

How do you put a giant telescope on the far side of the moon? The dish of the FAST radio telescope, for example, is made of thousands of triangular metal panels. But there is no way to get so many heavy panels to the moon and piece them together.

Instead, Saptarshi and his team are relying on a neat trick that takes advantage of the properties of the waves they hope to gather. Because radio waves are so long, the reflector does not have to be a solid layer of material.

❝ If you have a structure in which the holes are smaller than the wavelength by a factor of four, then the wave does not 'notice' and will reflect perfectly off that surface. ❞ —Saptarshi

Think of a grate over a drain. Water can flow through, but the grate repels larger objects. Rather than a solid material, therefore, LCRT can be made of a lightweight and flexible metal mesh.

A radio telescope reflector does not need to be a solid layer. The main reflector of the Arecibo telescope was made of gratelike panels.

Deployable structures

The vast, paper-thin mesh that will serve as LCRT's receiver presents another challenge. How do you fold it for launch so that it does not get tangled or damaged?

Frequently, scientists and **engineers** want to launch something into space that will not fit in a rocket's **payload bay.** They design *deployable structures* that can be folded to fit for launch and *deployed* (spread out) later.

The reflector on Saptarshi's LCRT will have a surface area of tens of thousands of square yards or meters. That is huge! But using fold patterns developed from origami, the reflector will be packed into a cylinder less than 10 meters (30 feet) in diameter. Such fold patterns have already been tested and deployed on smaller scales. Using this folding, LCRT will fit within standard rocket payload bays.

> ❝ We are riding on the shoulders of a bunch of giants who have designed very exquisite fold patterns that have already been deployed in space. There are many other mission concepts that also use these fold patterns. ❞
> —Saptarshi

> "We have a tensioned structure of interconnected wires that is holding up the much finer mesh." —Saptarshi

Upon deployment, the reflector mesh will be too thin and fragile to support itself. Therefore, Saptarshi's telescope will be mounted on a web of thicker support cables. The cables will be made of a material called carbon fiber, known for its high strength and light weight.

An engineer experiments with deployable structures at NASA's Jet Propulsion Laboratory (JPL).

Big idea: Harpoons and weights

LCRT will also deploy its supporting web of cables in a novel way. A **lander** will touch down in the center of the target crater. It will then fire 16 harpoons to the crater rim. Each harpoon will have anchors that dig into the lunar **regolith** like grappling hooks. (Regolith is the thick layer of dust that covers the moon's surface.)

> The front of each anchor has hooks. Once these anchors start dragging on the lunar regolith, they start digging deeper into the regolith and they catch on to the regolith. —Saptarshi

The support web can then be winched up, with the anchors carrying all the weight.

As with a traditional telescope, the reflector must be the correct shape to gather the incoming waves and bounce them to a detector. Cables of uniform weight will not take on the exact shape needed.

Saptarshi and his team have devised a simple fix. Small weights will be installed at certain points throughout the support web to pull the reflector into the correct shape. The slight modification should be enough to focus the radio waves to the antenna.

Saptarshi's friend at JPL, Masahiro Ono (also a NIAC Fellow), has proposed another novel use for harpoons in the exploration of small **solar system** bodies. You can check out his work in *Out of This World: Asteroid-Harpooning Hitcher!*

Inventor feature:
Keep innovating

> How do you design a mission for observing a signal that you don't know anything about? —Saptarshi

The design of LCRT has changed over time with refinements in the scope of the project.

> While we were doing the analysis, initially we came up with the idea that we needed to build a 1-kilometer (0.6-mile) telescope because we were trying to observe wavelengths on the order of 100 meters (330 feet). —Saptarshi

Such a colossal telescope would have required a crater 3 to 5 kilometers (2 to 3 miles) in diameter. Firing harpoons over 1 kilometer is not reasonable. Therefore, Saptarshi and his team proposed a two-**lander** concept. One lander at the center of the crater would have served as the base. An additional lander at the crater rim would have deployed **rovers** to secure the support

Saptarshi's first mission design called for using rugged DuAxel rovers developed at JPL. The rover still may be used in other missions to the moon or Mars.

web. But, as the team's understanding of the kind of signals LCRT would look for changed, the team reevaluated its design.

> Once we started looking at the current understanding of our science, we realized that there's an upper limit to how big these wavelengths can go. —Saptarshi

To home in on hidden signals from the early universe, the reflector needed to be about 350 meters (1,100 feet) in diameter, not 1,000 meters (3,000 feet). This smaller design would fit in a crater small enough for harpoons to be used. Saptarshi could now pack the entire telescope system into a single lunar lander, vastly simplifying its deployment.

Inventor feature:
Photography and national parks

In his spare time, Saptarshi enjoys nature photography.

> I like capturing the essence of nature in digital photographs. —Saptarshi

Saptarshi takes many of his photographs while on hikes in the U.S. National Park system.

> I have gone to 29 national parks and want to visit all 63 before I retire. —Saptarshi

Saptarshi also enjoys reading biographies of scientists, **engineers,** and other innovators who have changed the way we look at the world.

> Learning from their histories and experiences is very empowering. I like to understand their ideas of how they think of research. —Saptarshi

Lunar challenges

Building on the moon's far side presents more technical challenges than just getting the materials off Earth and onto the surface.

Earth's **atmosphere** holds and distributes heat around our planet's surface. As a result, temperatures only cool down modestly during the night. Without an atmosphere, the moon undergoes severe changes in temperature. During the lunar day, the sun bakes the surface to a temperature of 260 °F (127 °C). In the lunar night, temperatures plunge to about −280 °F (−173 °C). Such extreme temperature changes can cause metal to expand, contract, and eventually break.

The carbon-fiber support web will be able to tolerate such changes. But, the reflector mesh itself must be metal. The mesh will have to be tuned to maintain the correct shape during the cold lunar night, when LCRT will make its observations.

The powdery **regolith** also presents challenges to more than just anchoring the dish.

> ❚❚ The lunar regolith is very glassy and angular and very fine. It has this property of entering all our robot motors and messing things up very quickly. ❚❚ —Saptarshi

Once it is deployed, LCRT will have no moving parts to avoid such malfunctions.

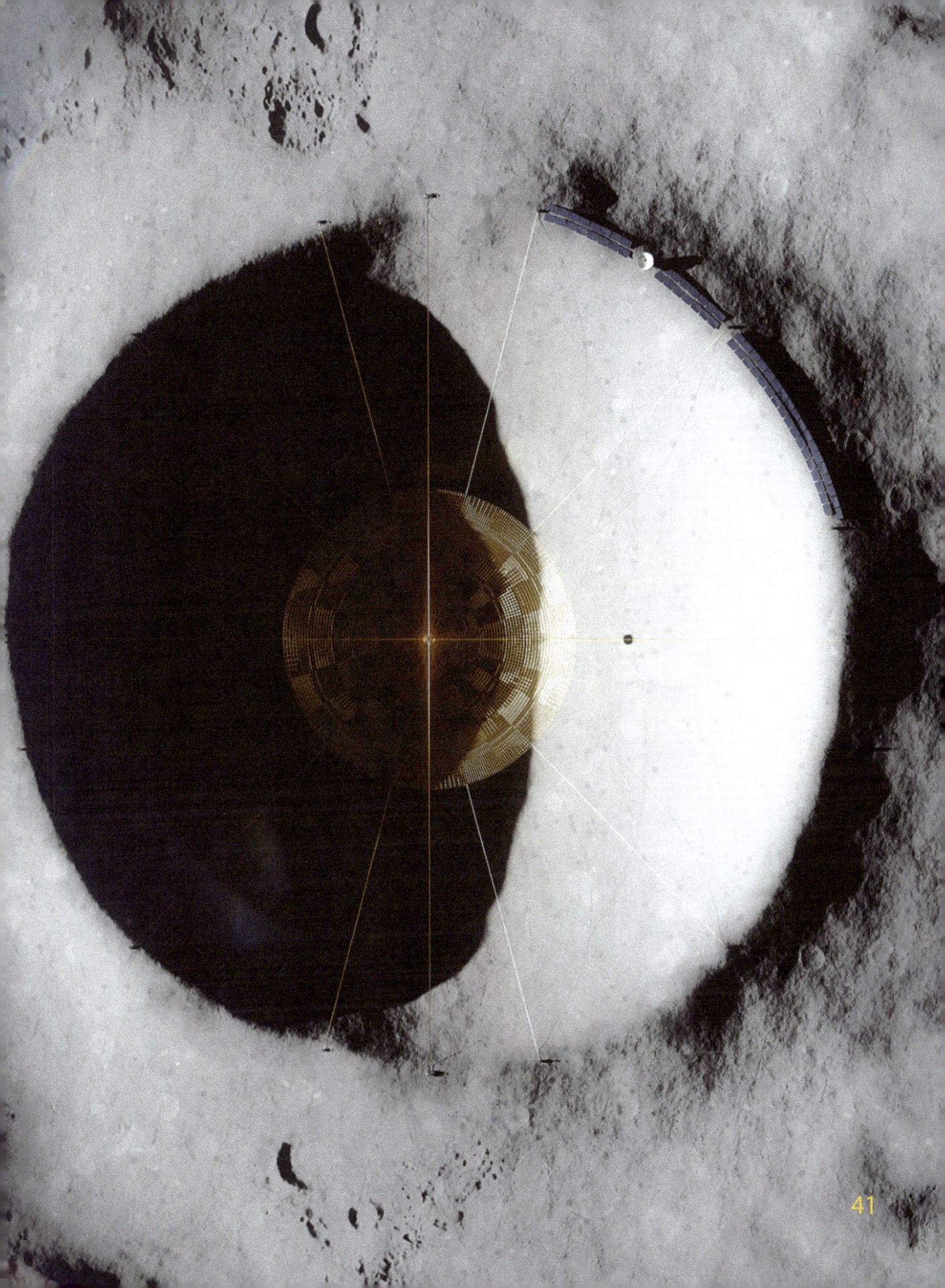

A lunar renaissance

The moon shields all wavelengths—including those used for radio communications in space. So, LCRT will need a way to relay its data back to Earth from the moon's far side. Saptarshi's team has worked on a small satellite for relaying signals. However, they do not envision that it will be necessary by the time the mission launches.

LCRT is being developed amid a *renaissance* (rebirth) in lunar exploration. Both the United States and China have plans to establish a permanent presence on the moon by the mid-2030's or so. Mission designs include **orbiting** space stations, trips to the surface, permanent bases, and the mining of resources. To support these ambitions, space agencies and private companies have developed plans to establish constellations of lunar satellites. Such constellations will orbit the moon and provide communications access to many or all parts of the surface. Saptarshi envisions LCRT using such a service to relay its data.

As human explorers gain a foothold on the moon, astronauts may even visit LCRT to perform maintenance or upgrades.

Saptarshi Bandyopadhyay and Haripriya Vaidehi Narayanan

Saptarshi lives in Los Angeles with his wife Haripriya, who is a scientist studying the immune system. They met at the Astronomy Olympiad, when they were both in high school. They built robots together in college, worked together on the Pratham satellite, and continue to explore the natural world together!

Glossary

astrophysicist a scientist who studies the physical traits of stars, galaxies, and other bodies.

atmosphere the mass of gases that surrounds a planet.

atom one of the most basic units of matter, consisting of a *nucleus* (core) of particles called *protons* and *neutrons* with tiny particles called *electrons* moving around the nucleus.

engineer a person who uses scientific principles to design structures, such as bridges and skyscrapers, machines, and all sorts of products.

gravitational pull also called gravitation or the force of gravity, the force of attraction that acts between all objects because of their mass. Because of gravitational pull, an object that is near Earth falls toward the surface of the planet. We experience this force on our bodies as our weight.

ionosphere the part of Earth's atmosphere, from about 34 to 190 miles (55 to 360 kilometers) above the surface, that has many electrically charged particles.

lander a spacecraft designed to land on a planet, moon, or other body in space.

orbit a looping path around an object in space; the condition of circling a massive object in space under the influence of the object's gravity.

orbiter a spacecraft designed to orbit a planet or other object in space.

payload bay the part of the rocket set aside for carrying cargo.

propulsion pushing something, such as a spacecraft.

radio astronomy a branch of astronomy dealing with the study of objects in space by means of radio waves that these objects give off.

regolith rock fragments, from large boulders to dust, covering solid rock.

rover a lander designed to move about for surface exploration.

sensor a device that detects heat, light, or some other phenomenon, producing an electric signal.

solar system the sun and everything that travels around it, including Earth and all the other planets and their moons.

wavelength the distance between two peaks of a wave.

Review and reflect

Now that you've finished reading about Saptarshi Bandyopadhyay, use these pages to think about his experiences and long-wavelength radio astronomy in new ways. As you work, reflect on the importance of creative problem solving, curiosity, and open-mindedness in life.

Complex problems and creative solutions

What is long-wavelength radio astronomy and why is it important?

What are some of the problems associated with long-wavelength radio astronomy?

How does Saptarshi Bandyopadhyay plan to overcome these challenges with the Lunar Crater Radio Telescope (LCRT)? What makes this solution so innovative?

Visit www.worldbook.com/resources to download sample answers, blank graphic organizers, and a rubric to evaluate writing.

Inspiration can come from anywhere!

Use a graphic organizer like the one below to map out your ideas. What ideas or experiences led to Saptarshi's innovative LCRT?

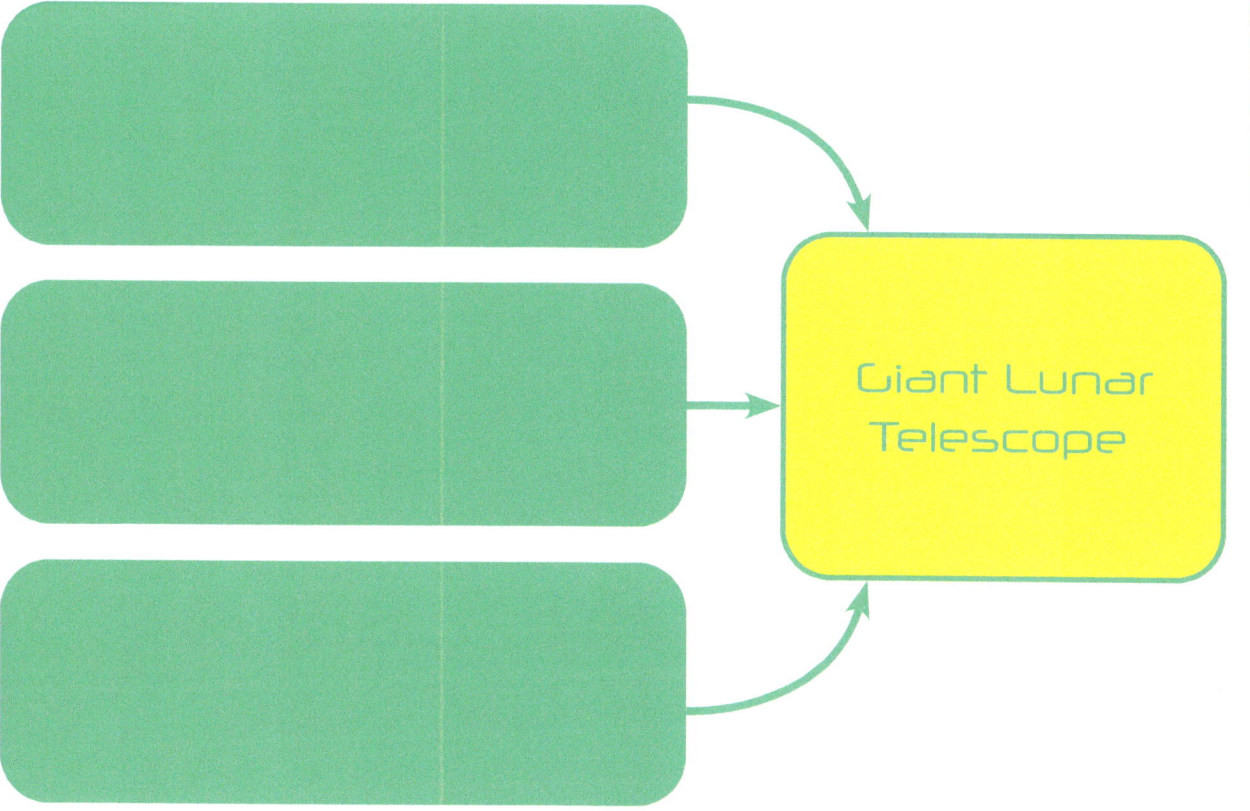

Write about it!

Think about Saptarshi's experiences in life and as a NIAC Fellow.

How have he and his team used a growth mindset to improve the LCRT design? Why might it be important to look beyond your initial concepts when searching for innovative solutions?

Index

A
Andromeda (galaxy), 8
Arecibo Observatory, 18, 26, 31
astrophysicists, 4, 8, 10
atmosphere, 4, 6, 23, 24, 40

B
Bandyopadhyay, Saptarshi, 6-7, 44; boyhood, 14; education, 28-29; Giant Metrewave Radio Telescope (GMRT), 14; hobbies, 38; International Astronomy Olympiad (IAO), 14, 44
big bang, 10-11, 13, 20

C
carbon fiber, 33, 40
craters, 26, 34, 37

D
Dark Ages, 10-11, 20
dark energy, 11, 12-13
dark matter, 11, 12-13
deployable structures, 32-33

E
Earth, 24, 27, 40

F
Five-hundred-meter Aperture Spherical radio Telescope (FAST), 18, 26, 30

G
galaxy, 8
Giant Metrewave Radio Telescope (GMRT), 14, 16-17
gravitational pull, 12-13
gravity, 27
Green Bank Telescope, 18

H
harpoons, 34, 37
hydrogen atoms, 10-11, 20

I
Indian Institute of Technology (IIT) Bombay Student Satellite Program, 28
ionosphere, 22-23, 24

J
Jet Propulsion Laboratory (JPL), 7, 33

L
lander, 34, 36-37
light, 8, 12, 20, 27
light wave, 16
light-year, 8
Lunar Crater Radio Telescope (LCRT), 6, 13, 20, 24, 27, 30, 32, 36

M
mesh, 32-33, 40
Milky Way, 8
moon, 22, 24, 26-27, 40, 42-43

N
NASA Innovative Advanced Concepts program (NIAC), 7, 35
National Aeronautics and Space Administration (NASA), 7

O
Ono, Masahiro, 35

P
Pratham (satellite), 29, 44

R
radio astronomy, 16, 24
radio telescope, 16, 18, 24, 27. *See also* Arecibo Observatory; Five-hundred-meter Aperture Spherical radio Telescope (FAST); Giant Metrewave Radio Telescope (GMRT); Green Bank Telescope; Lunar Crater Radio Telescope (LCRT)
radio waves, 16, 20, 23, 30
redshift, 20
regolith, 34, 40
rover, 36-37

S
star: formation, 11; Proxima Centauri, 8; radio interference, 27; Sirius, 8; sun, 8

U
universe, 8, 10-11, 12-13, 20

W
wavelength, 15, 16-17, 20, 22-23, 26, 30, 36-37, 42

www.ingramcontent.com/pod-product-compliance
Lightning Source LLC
Chambersburg PA
CBHW060933170426
43194CB00023B/2953